Also by Charlotte Herman

Millie Cooper, 3B
Millie Cooper, Take a Chance
The House on Walenska Street
Summer on Thirteenth Street

Millie Cooper and Friends

Millie Cooper and Friends

By Charlotte Herman

Illustrated by Helen Cogancherry

Viking

VIKING
Published by the Penguin Group
Penguin Books USA Inc., 375 Hudson Street, New York, New York 10014, U.S.A.
Penguin Books Ltd, 27 Wrights Lane, London W8 5TZ, England
Penguin Books Australia Ltd, Ringwood, Victoria, Australia
Penguin Books Canada Ltd, 10 Alcorn Avenue, Toronto, Ontario, Canada M4V 3B2
Penguin Books (N.Z.) Ltd, 182-190 Wairau Road, Auckland 10, New Zealand

Penguin Books Ltd, Registered Offices: Harmondsworth, Middlesex, England

First published in 1995 by Viking, a division of Penguin Books USA Inc.

1   3   5   7   9   10   8   6   4   2

LIBRARY OF CONGRESS CATALOGING-IN-PUBLICATION DATA
Herman, Charlotte.
Millie Cooper and friends / by Charlotte Herman;
illustrated by Helen Cogancherry.   p.   cm.
Summary: Millie, a fourth-grader in 1947, struggles with her feelings
and choices when her best friend seems to prefer the company of a new classmate.
ISBN 0-670-86043-3
[1. Friendship—Fiction.   2. Schools—Fiction.]
I. Cogancherry, Helen, ill.   II. Title.
PZ7.H4313Md 1995 [Fic]—dc20   95-18324   CIP   AC

Printed in U.S.A.
Set in New Century Schoolbook

to the newlyweds,
Karen and David,
with love

Contents

My thanks to Phil Chavin for sharing his memories of the cough drop, George Washington, and the Lawson Found; to Anita Sidler for the colored socks; and to Arlene Erlbach for the new girl in school.

1

School Supplies

Millie Cooper spread out her new school supplies on the dining-room table. She had her blue canvas three-ring notebook, two blocks of notebook paper—lined and unlined—a ruler, a box of reinforcements that looked like tiny white doughnuts, and a box of twenty-four Crayola crayons. There was a pencil box that held her pencils, a pencil sharpener, a combination ink and pencil eraser, a fountain pen, and an ink blotter from the John Hancock Life Insurance Company.

Best of all was her red-and-blue plaid schoolbag that came with both a shoulder strap and a handle so Millie could either wear it or carry it.

Everything was there and ready for tomorrow, her first day of fourth grade at the Victor Lawson Elementary School.

Millie glided her hand across the notebook paper. She rolled the pen and pencils between her hands and let her fingers play with the reinforcements inside the box. Then she ran them over the tips of the crayons.

How she loved the feel of new school supplies! And the smell, especially the smell of the crayons.

Nothing smelled as wonderful as a brand-new box of crayons.

Except maybe supper.

"Oh, I just can't wait for school tomorrow," Millie said later when she sat down to eat. "It's going to be the best year ever. For one thing, no more Miss Brennan. For fourth grade we get Miss O'Hara. The best teacher in the whole school. And maybe even the best teacher in the whole city of Chicago." She took a bite of liver and onions and chewed happily.

"I had Miss O'Hara when I was in fourth grade," said Millie's father. "Everyone liked her even way back then."

It was hard for Millie to imagine that her father or mother were ever in fourth grade. But she knew it was true. She had seen their class pictures.

"She's pretty old now," said Millie, picturing Miss O'Hara, with her thick white hair and rather large nose and thinking that she looked a little like George Washington. "But everyone says she's nice and kind. Strict but fair." Unlike Miss Brennan, who was strict and unfair, she thought. "Even her name has a nice sound to it. Kathryn O'Hara."

"Did you ever notice," said Millie's father to her mother, "that practically all teachers are named Kathryn? Or Margaret. Every teacher I ever had was either a Kathryn or a Margaret."

"I once had an Evelyn," said Millie's mother. "Evelyn Shapiro."

"Miss Brennan is a Margaret," said Millie. "I

2

remember the first time I saw *M. Brennan* written on my report card. I thought the *M* stood for Miss. Miss Brennan."

"Lots of teachers just use their initials," said Mrs. Cooper. "They don't want kids to know their first names."

"They don't want kids to think they even *have* first names," said Mr. Cooper.

"I found out that Miss Brennan had a first name by accident," said Millie. "Once I heard her talking to a substitute teacher in the hallway, and Miss Brennan said, 'How do you do? My name is Margaret Brennan. If the kids give you any trouble, let me know.'" She helped herself to another serving of liver and onions.

"You must really be looking forward to school," said Mrs. Cooper. "You never eat this much unless you're happy about something."

"I always look forward to the first day of school," said Millie. "And that's what I don't understand. How come I can't wait for school to start in September, and then I can't wait for it to end in June?"

She thought of that day this past June when she and her best friend, Sandy Feinman, had run out of the school laughing and screaming, and singing:

> *School's out, school's out,*
> *Teacher let the monkeys out.*
> *One went in, and one went out,*
> *And one went in the teacher's mouth.*

They had felt so free, as if they were birds being let out of a cage.

"It's all about change," said Millie's mother. "People need a change from routine sometimes."

"Sort of like going on a trip," said Millie's father. "You can't wait to leave and then you can't wait to come home again."

It had been like that for Millie this summer. She had felt that she couldn't wait to go to South Haven, in Michigan, where they had rented a cottage for two weeks. But there had been no kids her age to play with, and after a week of swimming and Bingo, she had been ready to go home.

Millie helped her mother clear the table while her father went into the dining room to listen to the radio. Ordinarily, Millie would have gone in later to listen with him. There were some good programs on. She especially liked *Fibber McGee and Molly*. She liked the part when Fibber opened the door to his overcrowded closet and all kinds of junk came crashing down.

But this evening, she went straight to her room to get her clothes ready for school. She took out her brand-new yellow sweater from her drawer, and her new yellow-and-green plaid skirt from her closet. Millie always wore plaid on the first day of school. Not that she especially liked plaid. But it seemed almost like a rule that on the first day of school you wore plaid.

The skirt and sweater looked just like the outfit

the girl on the cover of her *Polly Pigtails* magazine was wearing. She draped them neatly over the back of a chair. The girl was also wearing brown oxfords and white bobby socks—just like the new shoes and socks that Millie had. She placed them on the floor next to her bed.

Millie's school supplies were safely packed away in her schoolbag, and she opened it to take another look at them. Then she closed it up and set it next to the chair.

Mrs. Cooper came in and sat down on the edge of Millie's bed. Millie saw that she looked all misty-eyed, the way she looked every year at this time. First days of school were hard on her mother.

"You're growing up so fast," said Mrs. Cooper, picking pieces of chenille off the bedspread. "It seems like just yesterday that I took you by the hand to kindergarten, and here you are, going into fourth grade already."

"Fourth grade is going to be perfect," said Millie as she sat down next to her mother. "Room 212 has bigger desks. And sliding blackboards with the cloakroom behind them where you hang your wraps. And now we get to use the big gym instead of the little gym. I can hardly wait."

Her mother left, and Millie soaked in her Lavender Blue bubble bath, closing her eyes and humming the tune to "A Pretty Girl Is Like a Melody," the song she had heard at a Marshall Field's back-to-school fashion show. With her eyes still closed, she made her usual first-day-of-school promises. From this day forward, she would go to sleep early and get up early so she wouldn't have to rush to school. She would get her clothes ready the night before so she wouldn't have to make a frantic search in the morning for something to wear. She would give her mother plenty of time to braid her hair, and she would keep her school supplies neat.

And even though she made these same promises every year, this year she would keep her promises.

Because this year was different.

This year was going to be the best.

2

Tea for Two

In the morning, with her plaid schoolbag slung over her shoulder, and looking as much like Polly Pigtails as possible, Millie set out for school. She was going to meet Sandy Feinman on the corner of Thirteenth Street and Millard. They had been meeting there and walking to school together almost every morning since first grade.

The moment Millie spotted Sandy, she ran toward her. And soon they were hugging and squealing, as if they hadn't seen each other in years—instead of just since yesterday, when they went to buy their school supplies at Woolworth's on Roosevelt Road. Today Sandy was wearing plaid, too.

"This is going to be such a swell year," said Sandy as they started walking to Lawson School. "Do you realize that, as of today, we're entering the middle grades? No more milk and cookies for us. And, as fourth graders, we might even get to serve them to the kindergarten."

"I heard that Miss O'Hara takes her classes on lots of field trips," said Millie, giving a little skip. "She takes them to places like the Field Museum and the Shedd Aquarium. And she makes learning a positive experience." She had heard her father telling her mother that last night, just before she fell asleep.

They hugged and squealed again, and ran the rest of the way to school.

The playground was filled with kids. The boys were shouting and chasing each other. Howard Hall and O.C. Goodwin, the class troublemakers, were taking turns pounding each other on the arms.

On the other side of the playground, the girls were swinging, jumping rope, or just talking. Millie and Sandy waved to Marlene Kaufman and Angela Moretti to get their attention. But Marlene and Angela were too busy sneaking glimpses at the boys to notice them.

"Let's be the first ones to line up," said Millie, heading toward the entrance marked GIRLS. "Then we'll be the first ones in the room."

"We'll grab two seats together right up in front," Sandy added.

"Far away from O.C. Goodwin and Howard Hall." Millie still couldn't quite forgive O.C. for threatening to knock her block off when they were in third grade. And just because Millie was forced to write his name on the blackboard for talking.

"Don't forget," Sandy said after a while. "You're coming to my house for lunch."

"I can't wait. Will we be spilling tea?"

"Of course," said Sandy.

Millie loved going to Sandy's house for lunch. No one was home there, because both of Sandy's parents worked, and they had the house to themselves. And since they had stumbled upon their tea-spilling game, it was more fun than ever.

The first bell rang, and girls from all sides came running to line up. Millie could tell that everybody was excited about coming to school today. Never did the students look as fresh and clean as they did on the first day, with their hair neatly combed and their shoes polished and shiny. Millie had noticed earlier that even the boys looked clean.

As soon as the second bell rang, the doors opened. Two teachers were there to welcome them to their first day of school. One of them was the eighth-grade teacher, Miss Cook, otherwise known as "the Cough Drop" because she always smelled of Smith Brothers cough drops. The other was Miss Ryan, the gym teacher, who made all the kids in her classes wear green on St. Patrick's Day.

"All right, now, no pushing or shoving," said the Cough Drop. "Walk, don't run, and there is to be *no* talking!"

Millie and Sandy made their way up to the second floor as fast as they could without running. As they skipped up the stairs, Sandy said, "Hey, Millie, did you ever notice that Miss O'Hara kind of looks like George Washington?"

"She does," said Millie, giggling. "But better not let her hear you call her that." And they went running and laughing to room 212.

As Millie reached for the knob, the door opened and out walked George Washington. She smiled at them, and Millie's heart filled with joy.

"Good morning, girls," Miss O'Hara said. "Welcome back." Then she continued walking down the hall. Millie and Sandy watched her for a moment before turning back to room 212. As they were about to enter the room they both froze. Sitting at the teacher's desk, in Miss O'Hara's seat, was Miss Brennan.

What was Miss Brennan doing here? Had they accidentally gone to the wrong room? Millie and Sandy each took a step backward and looked up at the room number. Room 212 was clearly written on the small window above the door. What was going on?

Miss O'Hara always taught in room 212. Not only that, but it said room 212 on the back of Millie's third-grade report card. Right where it showed the number of times absent and the number of times tardy, Miss Brennan had written, *Promoted to Grade 4B, Room 212.*

Probably Miss Brennan and Miss O'Hara switched rooms at the last minute, Millie decided. Probably fourth grade was going to be in room 210 this year. But how could that be? What about the larger desks and sliding blackboards?

"Don't just stand there. Come in and take your seats." Miss Brennan was looking at them from her desk.

"But we're in fourth grade now," said Sandy.

"I know very well what grade you're in, Miss Feinman. Now take your seat. You too, Miss Cooper. Same seats as last year."

Reluctantly, Millie made her way to her seat in the third row, right behind Sandy, and waited for the rest of the kids to arrive. She watched the bewildered looks on their faces when they saw Miss Brennan, who was now standing in the doorway waiting for them. She watched them as they backed out to double-check the room number as she and Sandy had, and watched smiles turn into frowns as they entered the room to take their seats.

When everyone was seated, Miss Brennan took attendance, and walked to the front of the room.

"Boys and girls, you are no doubt wondering why I'm here instead of Miss O'Hara. The answer is simple. Miss O'Hara and I have decided to exchange classes this year. After all these years, we both felt the need for a change. Therefore, she'll be teaching third grade and I'll be teaching fourth."

Millie thought she heard a few groans coming from the back of the room.

"This should be an easy adjustment for all of us," Miss Brennan continued, "because I already know you. And you already know me."

More groans.

Millie placed her elbows on the desk—barely noticing that it was indeed larger—rested her chin in her hands, and stared at the back of Sandy's head.

Then she closed her eyes. It was almost like being in third grade again. She just couldn't bear the thought of having Miss Brennan for another whole year—another whole year of "Miss Cooper, do this" and "Miss Feinman, do that." Why couldn't she at least call kids by their first names? And what would happen if Miss Brennan decided to teach fifth grade next year? And then sixth grade the year after, and seventh, and eighth? What if she followed them all the way to high school?

When Marlene Kaufman, who was chosen to be one of the book monitors, passed out the fourth-grade books, Millie awoke from her nightmare. Normally, she loved getting new books. But this time, she didn't care about them. She didn't care about her new school supplies or the larger desks or sliding blackboards or looking like Polly Pigtails.

"What a revoltin' development this is," Sandy said, as they walked to her house for lunch.

Millie recognized the phrase from a radio program called *The Life of Riley*. William Bendix, the actor who played Riley, was always saying that, whenever something didn't turn out the way it was supposed to. And for the fourth graders at Lawson School, today hadn't turned out the way it was supposed to at all.

"They could've at least warned us," said Millie. "The school should've sent notes home. That way it wouldn't have been such a shock."

When they reached Sandy's apartment on Millard Avenue, Sandy opened the door with a key which she wore on a ribbon around her neck.

The apartment was empty and quiet, unlike Millie's apartment, which at lunchtime was always filled with the sounds of a running vacuum cleaner or of dishes being washed, and soap-opera voices coming from the radio. And where her mother waited with lunch and asked her how her day was going.

"Cheese sandwiches okay?" Sandy asked as they started toward the kitchen. "We can have them with chocolate milk."

"Sure," said Millie. Cheese sandwiches tasted best with chocolate malteds, but they were good with chocolate milk, too.

"And I have potato chips and pickles. And tea for dessert."

At the mention of the word *tea,* they both started giggling.

The tea-spilling game had begun in spring when Millie and Sandy were in grade 3A. They were eating lunch at Sandy's and drinking iced tea from teacups, pretending to be Big Ladies.

It was while they were being Big Ladies and holding their cups with their pinkies up in the air, that Millie lost her grip and accidentally spilled tea on the table.

The next thing Millie knew, Sandy was spilling tea on the table, too. On purpose. And she was laughing.

"Now it's your turn again," said Sandy. So Millie spilled more of her tea. And then Sandy spilled some. And before long, both of them were spilling tea. And both were laughing hysterically—laughing and spilling, spilling and laughing.

"Isn't this great?" Sandy had asked. "If my mother was home, we couldn't be doing this. I can do almost anything I want when I'm home by myself."

From that day on, whenever Millie ate lunch at Sandy's house, they spilled tea. "This will be a tradition for us," Sandy said one time. "The English drink tea at four; we will spill tea at noon. Even though it'll have to be iced tea. I'm not allowed to use the stove when my mother's not here."

Now they ate their sandwiches and drank their chocolate milk. They munched on potato chips and pickles.

Then Sandy and Millie cleared the table and brought their dishes to the sink. They brought their cups to the table, poured in the iced tea, and stirred in the sugar. They sipped the tea.

"This is so elegant," said Millie in a very ladylike voice.

"And delicious," said Sandy, and she turned her cup over and poured the tea out all over the table.

Millie took a few more sips, then turned her cup over and spilled out the rest of hers, too.

They burst out laughing and started singing

"Tea for Two." Then they did a little tap dance, with cereal bowls for hats, and a mop and broom for canes.

"Let's dress up as tap dancers for Halloween," Sandy said, as she twirled her broom.

"Or we could be witches," said Millie. "With that broom you look just like a witch."

"Yes, definitely," said Sandy. "We'll be witches for Halloween." And then they began cackling like witches. While they were cackling, Millie noticed the kitchen clock.

"Oh my gosh!" she cried out. "It's ten to one. We'll be late! And on our first day!"

They grabbed some dish towels and began wiping up the puddles on the table. Millie used the mop to wipe up the linoleum floor underneath.

"Good enough," said Sandy. "Let's go."

"But the table's still wet. And the floor, too."

"Leave it. I'll finish when I come home."

They ran all the way to school and made it to their seats just as the one o'clock tardy bell rang.

Miss Brennan spent a good part of the afternoon reminding the class over and over again that they were fourth graders now and she expected them to act like responsible people. It was the same old boring Miss Brennan talk. Millie couldn't keep herself from shifting around in her seat and looking out the window, wondering what was going on in Miss O'Hara's class. She stretched and yawned until finally Miss Brennan said, "Miss Cooper, please cover your mouth when you yawn. Nobody wants to see what's inside. Where are your manners?"

How Millie wished there was a way to get out of Miss Brennan's class. As she walked home that afternoon, she imagined her father coming home from work and announcing that they were going to move, and that Millie would have to transfer to another school. At another school, she would get a teacher as nice as Miss O'Hara.

But the more Millie thought about transferring, the more she didn't like the idea. If she transferred, she'd be in a new school with strange new kids.

By the time she reached her apartment, her new plaid schoolbag loaded with homework assignments, Millie knew she didn't want to transfer. She would just have to stick it out with Miss Brennan.

It looked like her perfect fourth grade was not going to be so perfect after all.

3

Colored Socks

The first few days of fourth grade were easy. Miss Brennan spent most of the time reviewing third-grade material with the class. They reviewed third-grade spelling and arithmetic and worked on their penmanship.

"Remember," said Miss Brennan, while Rochelle Liederman passed out the penmanship paper, "good penmanship is the key to success."

Sandy dropped a note over her shoulder and it fell onto Millie's desk. It read: *If good penmanship is the key to success, I'm in trouble.*

Millie wrote back: *No you're not. You can always become a doctor.*

And it was true, Millie thought. Once when she was sick and the doctor came to her house to examine her, he wrote out a prescription. Later that day, Millie's mother had said to her father, "How on earth can the druggist read this? The handwriting is awful."

And Millie's father had laughed and answered, "Don't you know? All doctors have terrible handwriting. It's probably a requirement for medical school."

Miss Brennan wrote letters, words, and sentences for the class to copy. Millie watched her at the board and decided that all teachers wrote the same way; smoothly, without making squeaking sounds with their chalk. Their letters were large and graceful and perfectly formed. And they wrote in nice straight lines across the board. Whenever Millie wrote on the board, her letters were small and uneven, and she always wrote on a slant. With handwriting like that, she could never become a teacher. Not that she wanted to, but still . . .

After penmanship, Miss Brennan gave a practice spelling test. It was while Millie was reminding herself about the "*i* before *e* except after *c*" rule, that the door opened and a girl walked in. She was followed by an eighth-grade boy. Everyone looked up from their papers.

The boy handed Miss Brennan a note and left. The girl remained close to the door, as if she were getting ready to make her escape. She had that neat and clean first-day-of-school look, even though it was already the third day and nobody in the class looked that way anymore. She was wearing a starched blue cotton dress and matching blue socks.

Millie had never seen blue socks before. She herself wore only white. So did all the other girls in her room.

Miss Brennan made an announcement. "Boys and girls, I want you to meet a new member of our class, Lenore Simon."

O.C. Goodwin and Howard Hall began reciting, "Simple Simon met a pieman going to the fair. . . ."

Miss Brennan shot them one of her cold stares, which shut them up right away.

Then she turned to Lenore and said, "Miss Simon, please tell the class where you're from."

Lenore stood straight and tall and said, "I'm from L.A.—that's Los Angeles—and we had to move to Chicago because my father relocated."

Millie wasn't exactly sure what *relocated* meant. Her mother had once dislocated her shoulder. Probably it meant the same thing. She wondered what part of Lenore's father was relocated. She couldn't figure out why they would move here, with all the wind and snow in the winter. Why didn't they just stay in L.A. until he healed?

When Lenore finished talking, Miss Brennan said, "Miss Simon, please find an empty seat and sit down."

Lenore took the last desk in Millie's row. Marlene Kaufman, who took her monitor job very seriously, leaped from her desk and brought Lenore a set of fourth-grade books.

Millie turned around to get another look at Lenore's blue socks, but she couldn't see them from where she sat.

"Face front, Miss Cooper," said Miss Brennan, and she went on to dictate more spelling words.

After spelling came a practice arithmetic test. Miss Brennan wrote ten multiplication problems on the board.

"Simple, simple," O.C. called out. This was immediately followed by a chorus—mostly boys—of "simple, simple."

"If these are so simple," said Miss Brennan, "you may all copy each of these problems ten more times for homework tonight."

"No fair," Millie and the other kids grumbled.

After the test, Miss Brennan called on Sandy to deliver a note to the office.

Whenever Miss Brennan chose a boy to go on an errand, she had him go alone. But when she chose a girl, she always said, "You may choose a partner."

Millie always chose Sandy, and Sandy always chose Millie. They loved walking through the quiet halls together, whispering and giggling and feeling special going on an important mission to the office. They walked slowly, stopping to peek into the other classrooms, and drinking at all the water fountains along the way, making the time away from class last as long as possible.

But now, instead of saying, "You may choose a partner," Miss Brennan told Sandy, "Take Miss Simon with you, so she will get to know the school."

"Can Millie come with us, too?" Sandy asked.

"How many people does it take to deliver a note?" Miss Brennan asked. "I should think that two are quite enough."

Sandy looked at Millie and shrugged. Millie, who was partway out of her seat and ready to go, sat back down.

A short time later, Sandy and Lenore returned.

Millie was happy to see them back so soon. That meant there had been no whispering and giggling between them, and no long stops at the drinking fountains.

But it would be quite a while before Millie and Sandy could go on any more errands together. She heard Miss Brennan tell Sandy, "Miss Feinman, for the next week, you will help our new pupil become familiar with the school and our classroom routine. You will be Miss Simon's classroom companion."

"Miss Simon's classroom companion." It sounded like the title of a magazine. Like *Woman's Home Companion*, the magazine her mother read sometimes. Millie could picture herself telling people, "I subscribe to two magazines: *Polly Pigtails* and *Miss Simon's Classroom Companion.*"

"And to make this arrangement easier," Miss Brennan went on, "I will ask Miss Cooper to change seats with Miss Simon."

Millie was jolted by the words. Change seats? What was Miss Brennan talking about? This was Millie's seat. She didn't want to change, to sit in the back of the room away from Sandy and right across from O.C. Goodwin.

Miss Brennan continued, "This will be just temporary, of course, but I would like each of you to take your personal belongings with you when you move. You may keep your books in your desks."

Miss Brennan stood in front of the room with her arms folded and waited. Millie emptied her desk of

her belongings and made the trek to the back of the room just as the blue socks came walking toward her. Millie thought she detected a smile on Lenore Simon's face.

Millie dropped down in her new seat. From across the aisle, she could hear O.C. Goodwin tease: "Aw, the Bobbsey Twins are being separated. Boo hoo."

Another revoltin' development, thought Millie. She would have to do something about this. But what? Maybe she could get her mother to write a note saying that Millie couldn't see the blackboard from way back here. But then Miss Brennan would probably tell Millie to get glasses. Millie hoped the week would go by quickly.

The following Monday, Lenore wore yellow: a yellow dress and matching yellow socks. She wore green the next day, and red the day after that. For a whole week, she wore socks that matched her dresses or skirts. Millie wondered how long it would take before she ran out of colors.

And for a whole week, Sandy showed Lenore how to find the girls' bathroom, which was way down in the basement, and the dreary lunchroom, which was also in the basement, and where the nurse's office and teachers' lounge were.

She explained the book-report chart, showing how each kid's tiny airplane climbed higher and higher on the chart every time he or she turned in a book report, and she showed her how to work on the individual reading units.

There was no chance for Millie and Sandy to pass notes back and forth, or test each other with arithmetic flash cards, knowing that they shared a secret from Miss Brennan: if you held the card up in front of the window, you could look through to the other side and see the answer backward.

Millie couldn't wait for the week to be over so everything could get back to normal, to the way it had been for her and Sandy before Lenore Simon and her matching socks became part of their fourth-grade class.

4

Shirley Temple

Millie loved Shirley Temple. The child movie star was cute and funny. She could act, sing, and tap dance. Every once in a while, a theater in the neighborhood showed an old Shirley Temple movie, and Millie always went to see it. In most of her movies, Shirley played an orphan all alone in the world until kind people came along and took her into their hearts and loved her.

So Millie was thrilled when Miss Brennan announced that the Lawndale Theater was featuring Shirley Temple in a special children's film series this fall, and Lawson students were invited to attend—at a special discount.

Miss Brennan asked the paper monitors to hand out sheets with information about the series for the kids to bring home to their parents.

"It's every Saturday for eight weeks," Millie told her mother and father that evening. "All you have to do is sign the permission slip and give me the money, and I'll get a little booklet with the tickets already in it."

The thought of being a subscriber to a film series, with her very own booklet, made Millie feel very important and grown up.

"The first movie, this coming Saturday, is *Heidi*," said Millie. "And the last one is *Rebecca of Sunnybrook Farm*."

"I remember when *Rebecca of Sunnybrook Farm* came out," said Millie's mother. "Because it was 1938—the year you were born."

"Really?" asked Millie.

"Really," said her father. He signed the permission slip and gave Millie the money. "That spunky little girl was singing and dancing her little heart out before you were even born."

Millie thought about that. If Shirley Temple was a little girl in 1938, and this year was 1947, then Shirley would be almost grown up by now.

The next morning, Millie ran all the way to the corner to meet Sandy. Her hair hung loose, except for where it was pinned back with barrettes, and it was flying in the wind.

"You're late, Miss Cooper!" Sandy called to her.

"Sorry," said Millie. She wound down to a walk. "But I woke up late and then I couldn't find anything to wear. Did you bring your permission slip and money?"

"Yeah, did you?"

Millie patted her schoolbag to show that they were inside.

"I can't wait for Saturday," said Sandy. "I love *Heidi*."

"So do I," said Millie. "It makes me cry. We'll have to remember to bring hankies."

Most of the kids, including Lenore Simon, returned their permission slips and received their booklets of tickets. Some of the kids had forgotten and said they would bring them tomorrow.

O.C. Goodwin and Howard Hall said they wouldn't be caught dead going to any girls' movies and in falsetto voices began singing, "On the Good Ship Lollipop."

Millie sat at her desk examining her booklet. On the cover it said *The Shirley Temple Film Festival*. A film festival! How wonderful! And inside were eight tickets, one for each film. Each ticket was a different color. *Heidi* was pink, *Rebecca of Sunnybrook Farm* was green, and there were red, white, blue, yellow, orange, and purple tickets in between. It was a beautiful booklet. And it made Millie even more eager to see the movies.

She couldn't wait for Saturday, when she and Sandy would sit next to each other and cry together. But because of Lenore Simon, the days were going by very slowly. Lenore had been in the class for more than a week, and Sandy was still being a Classroom Companion. It even looked like Sandy was becoming Miss Simon's Lifetime Companion. Wherever Sandy went, Lenore followed. Whatever Sandy did, Lenore wanted to do it with her.

Every time they lined up for dismissal or to go to the bathroom, Lenore was right there to grab

Sandy's hand before Millie even had a chance to get out of her seat.

She was right there on gym day when they were getting ready to line up. And in the gym, she hurried to pick Sandy as a partner the minute the teacher told the class that they had to work in pairs to do sit-ups and rope-climbing.

Sandy shot Millie a "what can I do?" look, and Millie was left to be partners with Myra Glass, a girl who had stringy hair and smelled like she never took a bath.

If all that wasn't bad enough, it seemed that Lenore was also reserving Sandy for the future. "Whenever we need partners in gym, I want you to be mine," she told Sandy.

Millie was angry. And hurt. Who did Lenore think she was, anyway? Millie was sorry that Lenore's father was relocated. And she hoped he was getting better. But Millie and Sandy had been friends for years, and in just a short time, Lenore Simon had taken over. Millie was angry with Sandy, too, for going along with Lenore. It was as if she didn't even care about Millie. Millie didn't know who to be more angry with, Lenore Simon or Sandy Feinman. Or Miss Brennan. She hadn't changed their seats back yet.

When school was over on Thursday, Sandy and Lenore lined up together. So Millie lined up with Angela Moretti. Millie liked Angela. She was short and cute and they had a lot in common. They had both had their appendixes taken out. And they both loved to eat

hot dogs at Dave's. Still, the idea of Sandy and Lenore . . .

The line passed into the hallway. "I can't walk home with you today," Sandy whispered. "I'm going to introduce Lenore to the library."

The Douglas Park branch of the Chicago Public Library was right across from Lawson School. All Lenore had to do was cross the street.

"Why can't she meet the library by herself?" Millie whispered back.

"She wants me to keep her company," Sandy said. And she and Lenore were out the door.

"Her socks are interesting," said Angela, as Millie watched the two girls run down the street.

"I like white," said Millie, wondering why Sandy hadn't asked her to go with them.

Millie walked Angela to her apartment building, which was on Thirteenth Street, right across from the school and next door to the Old Colony Bottling Company. Then she walked the rest of the way alone.

As soon as Millie stepped into her apartment, her mother took one look at her and said, "I've just baked a delicious chocolate cake. Why don't you have a piece, with a glass of milk. I'll join you."

"Chocolate cake's my favorite," said Millie.

"I know," said Mrs. Cooper, following her into the kitchen.

The kitchen was a warm, cozy place where wonderful cooking and baking smells lingered, and Millie felt better just being there.

Millie helped herself to a piece of cake and took a mouthful. "Mmm, delicious." She washed it down with some milk.

"I'm glad you didn't bake adult cake," said Millie, using the back of her hand as a napkin. Adult cake was what Millie called the coffee cakes and carrot cakes that her mother served to company.

"I'll let you in on a little secret," her mother said. "I like chocolate cake better than adult cake, too."

"I know," said Millie.

Mrs. Cooper cut a piece of cake and ate it without a fork—the way Millie did.

"Now, why don't you tell me what's wrong?"

Millie looked up at her mother. How could she tell that something was wrong? How could she always tell?

Millie lowered her eyes and played with the crumbs on her plate.

"I don't think Sandy likes me as much anymore. I think she likes this new girl, Lenore Simon, better."

Mrs. Cooper moved her chair closer to Millie. "Oh, honey, I'm sure that's not true. You and Sandy have been such good friends for so long. You two didn't get into an argument, did you?"

Millie shook her head.

"Then why do you think she likes the new girl better?"

Millie looked up at her mother and told her all about Sandy and Lenore.

"They're always partners for lining up and gym,

33

and they spend all their time together and go places together, like the library. Lenore gets in the way and Sandy lets her. And Miss Brennan changed my seat and didn't change it back, even though it was supposed to be just temporary."

"I'm sure *all* of this is just temporary," said Mrs. Cooper. "I'll bet that once Lenore makes friends with the rest of the class, she won't stick so close to Sandy."

Millie wished that Lenore would hurry up and make friends with the rest of the class, so she could have Sandy back the way it was.

On Friday Millie and Sandy walked home together after school, and Millie felt sure that her wish was starting to come true.

"Don't forget tomorrow," said Sandy, when they reached the corner of Thirteenth and Millard. "What time should I pick you up?"

"Come by at ten o'clock," said Millie. "The movie starts at ten-thirty and we want to get good seats. And then you can come to my house for lunch afterward."

"It's a deal," said Sandy.

"Great," said Millie, starting down Thirteenth Street. "See you tomorrow. And don't forget the ticket."

"See you tomorrow," Sandy called over her shoulder as she headed down Millard. "And don't forget the hankie."

5

At the Movies

"I'm all ready to go," said Millie. She tore the pink ticket out of her booklet. "All I need now is a hankie."

"I have just the thing," said Mrs. Cooper. She went into her room and came back out a few moments later.

"I used this when Olivia de Havilland lay dying in *Gone with the Wind*."

Millie took the hankie her mother held out to her. It was white and had tiny pink roses embroidered along the edges.

"Oh, Mama, it's beautiful. Thank you. And I won't use it to blow my nose. I'll only use it for tears."

Millie was checking her money to make sure she had enough for some Milk Duds or Jujubes when the doorbell rang.

"Oh, good. Sandy's right on time. See you later, Mama. And don't forget—Sandy's coming back here for lunch." Millie ran to open the door.

"You're right on ti—" She stopped short. Because standing next to Sandy was Lenore Simon.

Millie's heart sank. What was Lenore doing here? It was supposed to be Millie and Sandy's day. Their time together. Alone.

"Oh . . . hi," Millie managed to say in a weak voice.

"All set? Got everything?" asked Sandy.

"We should hurry so we can get aisle seats," said Lenore. "They're the best."

Millie noticed that Lenore was wearing the same blue dress and socks she had worn on that first day. She finally ran out of colors, Millie decided.

"I thought the seats in the middle of the row were the best," said Millie, as the three girls walked out of the building.

"The aisle is better because you can get in and out real easy if you want to buy some candy or go to the bathroom."

"What about the kids who have to climb over you when they have to go in or out?" asked Sandy.

Lenore just shrugged.

A line was already forming outside the Lawndale Theater, and the girls hurried to get into it.

"I didn't know Lenore was coming with us," Millie whispered to Sandy while they were waiting.

"She didn't want to go alone, so I invited her. It's okay, isn't it?"

Millie didn't even have a chance to answer—to tell her that no, it wasn't okay—because at that moment, the doors opened and all the kids began rushing in. They stopped just long enough to turn in their tickets and get the stubs back.

"Keep the stub," Sandy told Lenore. "They're giving away door prizes later."

"This is so exciting," said Lenore. "My first Chicago movie."

The aroma of popcorn filled the lobby, and lots of kids headed over to the candy counter.

"Let's get our seats first," said Millie, "and we'll come back here later."

With Millie in the lead, they made their way down the aisle until they found an almost empty row in the center of the theater. Millie was starting toward the center of the row, when Sandy called to her, "Millie, over here."

Millie looked back and saw that Sandy and Lenore had chosen seats near the aisle. What was wrong with Sandy? She and Millie always sat in the middle of the row whenever they went to the movies. Why was Lenore suddenly in charge?

Millie went over to where they were sitting and dropped down in the seat next to Lenore.

"If you give me some money, I'll go get the candy," said Sandy. "What do you want?"

"Jujubes, I guess," said Millie.

"I'll have Milk Duds," said Lenore. "But wait—I'll go with you." And like a shadow, Lenore followed right behind Sandy.

Millie slumped down in her seat and looked around the theater. She saw lots of familiar faces from Lawson School. Two of those faces belonged to O.C. Goodwin and Howard Hall, who looked like they were trying to hide in the back row.

When Sandy and Lenore came back with the

candy, Millie opened her box of Jujubes right away. She popped a couple of red ones into her mouth. She liked the taste of Jujubes. They were fun to eat. The only thing she didn't like was the way they got stuck in her teeth.

Little by little, the theater began to fill up. And as it did, it became noisier and noisier. Kids were running around and shouting and having popcorn fights.

Finally the manager of the theater walked up to the stage and announced over the microphone, "All right, boys and girls, settle down. We're going to have our drawing for the door prizes."

The kids cheered and whistled and stomped their feet.

"Today we'll be giving away a Flexible Flyer sled and a pair of ice skates—hockey skates if the winner is a boy, and figure skates if it's a girl."

"I would love to win the sled," said Lenore. "There was never a chance to go sledding in L.A. No snow."

Millie already had a Flexible Flyer. But she thought that a pair of figure skates would be wonderful. She imagined herself skating on the frozen pond in Douglas Park. Skating like Barbara Ann Scott, the famous world champion ice-skater.

The manager chose a girl to come up to the stage. He had her pick a ticket stub from a box he held out to her.

Millie looked at her own ticket stub with the number 511 written on it.

The manager took the ticket stub from the girl and announced, "The winner of the Flexible Flyer is number 613."

"I can't believe it!" Lenore shrieked. "I missed by one number. I have 513."

"Actually, you missed by a hundred numbers," Millie was happy to tell her.

A small boy who looked like he was in kindergarten won the sled. "The sled is ten times bigger than the kid," said Sandy. Even Millie had to laugh at that.

When it came time to choose the winning ticket for the skates, Millie held her breath. In her mind, she was still skating around the pond, wearing her white skates with tiny silver bells on the laces and silver blades that flashed in the sunlight. Actually Millie didn't even know how to ice skate. Still . . .

"And the winner of the skates is—number 151."

Millie let out a deep sigh. Not even close. A girl Millie recognized from Lawson School came up to claim her prize.

The drawing was over. The lights went out, and a Tom and Jerry cartoon appeared on the screen. The audience cheered again. After the cartoon came a newsreel of world events and a coming attraction of next week's Shirley Temple film, *Little Miss Broadway*.

Oh, I can't wait to see that, Millie said to herself, as she rolled some Jujubes around in her mouth.

Millie watched as a smiling Harry Truman announced that he was looking forward to being the first president to speak to the nation on television.

Finally *Heidi* was on. The audience was quiet for most of the movie. But the kids booed when the mean aunt kidnapped Heidi from her grandfather's mountain cabin in Switzerland. And they booed again when the governess tried to sell Heidi to the gypsies.

Throughout the movie, Shirley remained spunky. And throughout the movie, Millie found reasons to cry. But today she couldn't cry with Sandy, because she wasn't sitting next to her. And she didn't feel like crying with Lenore or holding her hand at the sad parts. So she cried alone.

She cried when Heidi pleaded to be taken back to her grandfather, and when the crippled girl named Clara got out of her wheelchair and walked—even though everyone had said she would never walk again. She cried at the end of the movie when Shirley Temple, who seemed to be looking right at Millie, said, "And please make every little boy and girl in the world as happy as I am."

Millie wiped her eyes with the damp hankie. She wondered if her mother had cried this much when she watched Olivia de Havilland die.

When the movie was over, the audience burst into applause and the lights went on again. Millie, Sandy, and Lenore left the theater singing the song Shirley sang about her new wooden shoes.

"That movie was so good," said Millie, squinting in the daylight.

She spotted O.C. Goodwin and Howard Hall

41

again. This time it looked like they were trying to get lost in the crowd. Millie pointed them out to Sandy and Lenore.

"Oh, Howard!" Lenore called out. The three girls moved toward them. "Hi, O.C. Imagine meeting you two here. How did you like the movie?"

Millie was shocked. She had never heard anyone tease O.C. or Howard before. It was always best to ignore them or they'd threaten to knock your block off.

"Aw, we didn't even watch it," said O.C. "We came for the cartoon."

"And the prizes," said Howard. "I wanted to win the sled."

"I'll bet you did," said Lenore, and the girls ran away, laughing.

"Do you want to come to my house for lunch?" Millie asked Lenore as they walked toward Millie's apartment building. She hoped Lenore would say no.

"Lunch? Didn't Sandy tell you? My father wants to take the three of us for corned beef sandwiches at Carl's."

"Oops, I meant to call you last night to tell you," said Sandy.

"But we had plans," said Millie.

"I know," Sandy answered. "But corned beef sandwiches at Carl's. With chocolate phosphates. Oh, Millie, please come with us."

Millie loved corned beef sandwiches. And she loved eating at Carl's deli. But that wasn't the point. The point was that Lenore was running things again.

"I don't think I'll be able to," Millie said. "My mother and I might decide to go downtown."

Sandy and Lenore dropped Millie off at her building. Then they continued on. Sandy and Lenore. Sticking together like two Jujubes.

6

Oatmeal Cookies

It was tops season at Lawson School. Spring saw yo-yos and rubber balls. But the wooden tops came out in the fall. Almost everyone—especially the boys—was playing with them at school.

One day Millie brought her red-and-blue striped top to school. She and Sandy spent their recess spinning their tops together on the concrete walk of the playground.

"I wish I could throw it overhand the way the boys do," said Millie, winding the string around her top. "I can only do it underhand." She threw her top out, quickly pulled back on the string and watched with delight as the top spun strong and steady.

Sandy threw her top out, too, and soon their two tops were spinning side by side.

Lenore Simon stood nearby and watched them. She didn't have a top, and Millie was glad. For once, Millie and Sandy could do something together without Lenore getting in the way.

But the next day, Lenore came to school with a top, and there she was, getting right in the middle again.

It wasn't that Millie disliked Lenore. Lenore

was okay. And she knew lots of things about Hollywood. She knew where the movie stars lived, and she had even seen some of them in person. Like Margaret O'Brien.

"I saw her walking on the lot at MGM Studios," Lenore told her and Sandy the following Saturday. They had just tap-danced out of the movie theater after seeing Shirley Temple in *Little Miss Broadway*.

"How did you get to go on the lot?" asked Millie, who wasn't exactly sure what "the lot" was.

"My father knows some important people in Hollywood. One of them took us there. I bet if I'd stayed in L.A., I might have become a movie star, too. In fact, I'm thinking of taking tap dancing lessons. And maybe even modeling lessons at the Patricia Stevens Modeling School, so I can gain poise and self-confidence, in case I ever move back to L.A."

And of course, Lenore knew lots about matching colored socks. She never once wore white.

So the problem was not Lenore herself, Millie decided. And if she had met her anywhere else but in Miss Brennan's fourth grade class, there wouldn't even be a problem.

No, the problem was that Lenore was making things different between Millie and Sandy. And that's what worried her. What if Sandy really liked Lenore better? What if one day Sandy decided to make Lenore her best friend? Where would that leave Millie?

Millie's mother had said that all Lenore needed was time to make more friends and she wouldn't stick so close to Sandy. So far that hadn't happened. Lenore hadn't even tried to make any friends. She stuck to Sandy like glue. If only Millie could do something about that. If only Miss Brennan would change her seat back.

Millie had always made it a practice to stay out of Miss Brennan's way whenever possible. But on the Monday after they had seen *Little Miss Broadway*, seen again how spunky Shirley Temple was, Millie decided that she could be spunky, too.

Just before music at the end of the day, Millie took in a deep breath, got out of her seat and walked up to Miss Brennan's desk.

"Miss Brennan," said Millie, "can I please move back to my own seat?"

Miss Brennan looked up at Millie and thought for a while. "Well, I don't see any reason for changing at this point."

"But you said—"

"Please sit down, Miss Cooper. We'll be starting music soon."

Maybe *spunky* worked for Shirley Temple, thought Millie as she headed for her miserable seat at the back of the room. But it sure wasn't working for her.

Millie's heart was not in the song Miss Brennan started to play on the piano. It was a song about being good neighbors with our friends across the Rio Grande. The class began singing, but Millie didn't sing along. She didn't feel like it. She didn't feel like being a good neighbor. But so what? Miss Brennan wasn't being a good one either. Besides, Miss Brennan wouldn't know if Millie was singing or not. She couldn't see her way back here. The class went on to sing about belonging to one big happy family. Millie would never want Miss Brennan in her family.

Miss Brennan stopped playing. "Remember, class, when we get to the part about extending a helping hand across the Rio Grande, I want you to shake hands with your neighbors to the right of you, to the left of you, to the front and to the back of you."

"What if no one sits in back of you?" O.C. called out. "Or in front of you?"

"Or to the right of you or the left of you?" Howard Hall added.

"We've been through all this in third grade," said Miss Brennan. "In cases like those, just shake hands with the air."

Miss Brennan went back to her playing and the kids—except for Millie—went back to their singing and extending their helping hands.

Everyone started shaking hands with each other. Millie looked up the row and saw Sandy and Lenore shaking hands and laughing. O.C. Goodwin held out his hand to Millie and she was forced to take it, afraid that he probably had a piece of chewed-up gum hidden there waiting for her. But to her surprise there was nothing in it. The problem was that he kept shaking her hand and wouldn't let go.

"Miss Cooper and Mr. Goodwin," came Miss

Brennan's voice from the front of the room, "you can stop shaking hands now. The song is over."

The next day during recess, while Millie, Sandy, and Lenore were spinning their tops, Angela Moretti, who was standing next to Millie, said, "The three of you sure are getting to be good friends."

"Yes," said Millie. "But it would be nice if Lenore got to know some of the other kids better. Like you, for instance."

"I could invite her to my house after school tomorrow. Marlene and I are going to bake cookies."

"That's a terrific idea," said Millie, feeling her spirits rise.

"Maybe you and Sandy can come, too."

"Maybe," said Millie, although she really wanted Lenore to go alone.

But when Angela asked Sandy and Lenore, Sandy said she was going to the dentist after school, and Lenore said that she and her mother were going to check out dance studios.

"Drat," said Millie under her breath. It was a word she had often seen in her comic books, and she had always wanted to try it out.

The next afternoon, Millie went to Angela's apartment to bake cookies. She had been there once before to visit, when Angela came home from the hospital after she had her appendix taken out.

Angela's grandmother, who sometimes stayed with her while her mother was at work, was there, too.

"Remember," she told Angela, "you call me when

you need to light the oven. You don't light it yourself. Remember what happened last time."

"What happened last time?" Millie asked, when she and Angela and Marlene went into the kitchen.

"Chicken feathers," said Marlene.

"Chicken feathers?"

"Yeah, the whole kitchen smelled like burnt chicken feathers," Marlene said, and she burst out laughing.

Angela started laughing, too. "We were getting ready to bake the cookies, and we needed to preheat the oven. So I turned on the oven and lit it with a match, the way my mother always does. But nothing happened. So I stuck my head in the oven to see what was wrong. The next thing I knew, pop! When I pulled my head out—"

"Burnt chicken feathers," Marlene finished for her.

"The whole top of my hair was singed."

By this time all three girls were laughing.

"And my mother said if I ever light the oven again, I'd have more than singed hair to worry about. She'd break my neck."

Angela was laughing so hard by now that she had to hold her stomach. Millie thought it was lucky that Angela had already had her appendix taken out, or it would have burst right there in her kitchen.

After everyone recovered, Angela said, "Okay, let's get started. Today we're going to bake oatmeal cookies."

"Big deal," said Marlene. "We always bake oatmeal cookies."

"Well, how else can I get rid of the oatmeal? If I don't use it up, I'll have to eat it for breakfast."

"You don't like oatmeal either?" Millie asked. Millie's mother often made hot oatmeal for Millie's breakfast—especially in the winter. She was a believer in hot cereal. But Millie didn't like it and always had to force it down.

"I only like it in cookies," said Angela.

"Me, too," said Millie. Now there were three things that Millie and Angela had in common: hot dogs at Dave's, no appendix, and not liking hot oatmeal.

Angela's grandmother preheated the oven, while the girls got all the ingredients together. They took turns sifting the flour and baking soda and stirring it into the mixture of eggs, butter, sugar, water, and oatmeal. Angela took a wooden spoon and held it in front of her mouth like a microphone.

"Mothers, do you want the best for your children? Of course you do. Then treat them to a fresh batch of oatmeal cookies every day."

Millie took the spoon from her and continued: "Ninety-five out of one hundred doctors—all with bad penmanship—agree that oatmeal cookies are more nutritious than hot oatmeal."

"So," Marlene went on, "be sure to get the whole family off to a good start with oatmeal cookies for breakfast every morning."

"Oatmeal batter is good for the skin too," said Angela, chasing Millie and Marlene around the kitchen with a finger full of batter. They all ran screaming and laughing, until Angela's grandmother chased them out and finished the baking herself. But she called them back in when the cookies were done.

They ate the warm cookies with cold apple cider. Millie thought about how much she liked Angela and Marlene. But she couldn't help it, she missed Sandy.

7

Changes

October came and autumn was in the air. Millie loved the change of seasons. She loved the way the leaves turned colors, from green to red and gold.

"This is what I like about Chicago," Millie said to Sandy on the way to school one morning. "Just when you get tired of one season, another is ready to take its place." She bent down to pick up some red maple leaves that had blown onto the sidewalk. Miss Brennan had asked the kids to bring in autumn leaves to decorate the classroom. Millie stuck them in her notebook, which by now had loose pages sticking out from all sides.

"Autumn is a whole new experience for Lenore," said Sandy as she collected some gold leaves that had fallen from an elm tree. "The seasons don't change in L.A. It's always summer."

"Boring," said Millie.

They passed a house where leaves and pumpkins were heaped on the front lawn.

"Look how pretty," said Millie. "It's like a scene on a calendar. You look at it and you know that fall is here."

"And pretty soon Halloween will be here. And the parade."

Every year, Lawson School held a parade in honor of Halloween. At lunchtime the kids dressed up in their costumes, and later in the afternoon, they went outside and paraded around the school. After the parade, each classroom held a party that lasted for the rest of the day.

"We should start planning our witch costumes," said Millie.

"That's what I wanted to talk to you about," said Sandy. "Lenore had this great idea. She told me about it when she came to my house for lunch yesterday. She thought—"

"Lenore went to your house for lunch?"

"Yeah, well, we didn't plan it. You see, when she got home she remembered that she didn't have a key. She never needed one before because her mother's always there. But yesterday her mother went downtown and . . ."

Millie barely heard the words. Images of Sandy and Lenore laughing and spilling tea filled her mind.

"So she came to my house," Sandy continued, "and asked if she could have lunch with me. Anyway, she had this idea that we could go as the three blind mice."

Laughing and spilling tea. Dancing and singing "Tea for Two"—without her.

"No," said Millie. "You and I were planning to go as witches. It was all decided."

"But her mother sews. She can make the costumes, with ears and tails and whiskers. They'll be so cute."

"No," Millie repeated. "I'm not going to school as a mouse. You and Lenore can if you want to. But I'm not."

"Who ever heard of two blind mice?" Sandy asked. "It only works with three. But it's no big deal. I'll tell Lenore that we'll all go as witches."

Millie didn't really care if they went as witches or not. That's not what mattered. What mattered was Lenore's changing things again.

After school Millie curled up on the kitchen window seat and leafed through her latest issue of *Polly Pigtails*. It was the Halloween issue and was filled with ideas for costumes, parties, and decorations.

"Mama," said Millie, "can you help me make a witch costume for the parade?"

Her mother looked up from a crossword puzzle she was working on at the table.

"You and Sandy are going as witches this year?"

"And Lenore."

"Oh, the three of you. How nice."

"Not so nice. And it wasn't my idea." Millie closed the magazine and placed it on her lap. "Even Halloween is different this year. Everything's different because of Lenore."

Mrs. Cooper got up from the table and Millie made room for her on the window seat.

"Oh, honey. Of course things are different. You're growing up. Nothing stays the same. Change is a part of life." Her mother smoothed back the wisps of hair that had fallen onto Millie's face.

"Think of all the new friends you'll be making as you grow older. All the new people who will come into your life. You've got to make room for them. You can have more than one good friend, you know."

Millie knew that she could have more than one good friend. She already had Angela and Marlene as friends. But her friendship with Sandy was different. Sandy was her *best* friend. It hurt so much to think that maybe Sandy liked Lenore better.

Millie picked up her magazine and rolled it up like a telescope. "Mama," she said looking through it, "remember you said that Lenore wouldn't depend on Sandy so much once she got to meet new people? Well, so far it hasn't happened. And I tried to do something about it. I even had Angela and Marlene come to some of the Shirley Temple movies with us. But that didn't work, either. Lenore still sat next to Sandy and hardly paid any attention to the rest of us. It's hopeless."

"Maybe with Lenore it'll take a little longer," her mother said.

Millie unrolled her telescope and read the article about Halloween parties. Now that was an idea. A party. Maybe someone from her class would have a Halloween party and invite Lenore. And if no one else gave a party, maybe Millie could have one. Maybe there was hope after all.

8

Ready for Halloween

"I'm having a Halloween party, and I'm inviting all the girls in the class," Lenore told Millie and Sandy at recess one morning. "My mother wants me to make new friends."

"Oh, that's a wonderful idea," said Millie, trying not to show her excitement. It was more than she could have hoped for. "It's good to bring new people into your life."

Sandy gave Millie a strange look, but Lenore continued on as if she hadn't heard. "It'll be on Halloween, one week from today, at six o'clock. So save the date. We'll have hot dogs and candy and we'll bob for apples and play games and I'll give a prize for the best costume. I'm passing out the invitations at three o'clock today when school lets out."

To her surprise, Millie was getting very excited about the party. It sounded like it would be lots of fun, and she was looking forward to going.

When they came in from recess, Miss Brennan was not in the doorway waiting to tell the kids to stop running and walk quietly into the room. She wasn't at her desk either.

As the class sat down to wait for her, Millie

heard giggling. She saw the kids pointing to the sliding blackboards. Two of the boards were all the way down to the floor. But one board had a large space under it. And in that space were feet—Miss Brennan's feet. Millie could recognize the thick black shoes and the hem of Miss Brennan's dress.

"What's she doing in there?" Rochelle asked in a whisper.

"Undressing," said Howard Hall.

"Going to the bathroom," said O.C. Goodwin.

The giggling grew louder. And just as it did, up went the blackboard and out walked Miss Brennan. She was carrying a coffee cup in one hand and a chair in the other.

"Big deal," said Marlene Kaufman. "She was just drinking her coffee."

Millie knew why Miss Brennan was hiding. Teachers didn't want kids to see them eating or drinking.

Just before three o'clock, Lenore went up to Miss Brennan's desk and asked if she could pass out her invitations.

"If everyone lines up quietly," Miss Brennan said in a voice loud enough for the whole class to hear, "and the line is nice and straight, you may pass them out. Otherwise you will have to wait until tomorrow."

At three, the class—even the boys—lined up quietly. The line was nice and straight. So Miss Brennan told Lenore that she could pass out the invitations. Lenore passed them out to all the girls. She even gave one to Miss Brennan, who seemed surprised to get one.

Millie was just as surprised. "You invited Miss Brennan?" she asked.

"Sure. In my last school kids invited teachers lots of times."

Millie hoped that Miss Brennan wouldn't come to the party. She probably wouldn't let them have any fun. She'd probably make them sit quietly with their hands folded.

"Hey, what about us?" O.C. Goodwin asked.

"I'm only inviting the girls," said Lenore.

"No fair," said Howard Hall, and all the boys started booing.

Miss Brennan placed her hands on her hips. "Perhaps you boys would like to have some extra long-division problems for homework."

The booing stopped, and Miss Brennan led the class out into the hall.

"Do you think Miss Brennan will come to Lenore's party?" Millie asked Sandy on the way home. She was looking at the invitation. The front of the card showed a broom-riding witch silhouetted against the moon, and the words:

Ghosts and goblins,
Witches mean,
Want you with them
For Halloween!

The date, place, and time of the party were written on the inside of the invitation.

"There's no chance she'll come to the party," said Sandy. "She wouldn't be around kids any more than she has to."

"I wonder what kind of costume she'll wear if she *does* come," Millie said.

"Little Red Riding Hood," said Sandy, laughing. "Picture a very large Miss Brennan in a red cape and hood."

"Or a very large Miss Brennan as a ballerina with a pink tutu." Millie stood up on her toes and did a pirouette. And before long, Millie and Sandy were dancing on their toes toward home, laughing and singing a melody from *Swan Lake*.

Millie's mother had taken Millie and Sandy to Orchestra Hall a few months ago to see *Swan Lake*. They climbed up the hundreds of steps to the gallery and had to take turns watching the ballet through binoculars. It had been worth the climb. The music and dancers were wonderful. But now, instead of picturing the ballet dancers, Millie imagined Miss Brennan in the pink tutu.

Millie spent the next few days getting her costume ready for Friday. The girls had decided that each of them would make her costume different. "It'll be more creative that way," Millie had said. And Sandy and Lenore had agreed.

Millie's mother gave her a long black skirt and

a black blouse. For a hat, Millie used a large sheet of black construction paper that she bought at Dennison's. She rolled it up in the shape of a cone and put paste along one of the edges so it wouldn't come apart. She cut out a ring of construction paper and used it as a brim for the hat.

She bought black licorice at the candy store and got some gum from her father, who always kept a pack in his jacket pocket. She was ready for Halloween.

On the morning of the parade, Miss Brennan spoke to the class about their costumes. "No long costumes to trip over. And if you must wear a mask, make sure the eye slits are cut wide enough for you to see clearly."

"Are you going to wear a costume?" O.C. asked.

"Don't be silly," said Miss Brennan. Millie pictured Miss Brennan in her pink tutu and had to stifle a giggle.

At noon Millie raced home to turn herself into a witch. She dressed up in the skirt and blouse and put on her hat. She chewed the gum and rolled it around between her hands until it was sticky. Then, looking in the mirror, she stuck the little ball on the end of her nose for a wart. She put some black licorice on her two front teeth so it would look like the teeth were missing.

Millie grinned at her reflection and cackled. "Mirror, mirror on the wall. Who's the fairest one of all?"

"You're an adorable witch," said her mother.

"A witch is not supposed to be adorable," said Millie. "She's supposed to be wicked."

"Then you're an adorable wicked witch." Her mother handed her a broom, a bag of cookies, and an apple.

"Heh, heh, heh, my poisoned apple. Thank you, my lovely." Millie headed for the door and almost tripped over her broom. "Oh . . . I almost forgot. I'll be home a little late. I'm going with Sandy and Lenore to Woolworth's after school to help pick out decorations for Lenore's party."

There was another witch waiting for Millie on the corner of Thirteenth and Millard, and she was waving to Millie. Millie rode her broom over to meet her.

Sandy was wearing a store-bought costume: a black witch's cape, a hat, and a mask. She was holding the broom she had used in the "Tea for Two" dance.

"Hello, dearie," said Millie. "Would you like a bite of this shiny red apple? Heh, heh, heh."

Sandy lifted up her mask, grabbed the apple, and took a bite out of it.

"Hey, I was only kidding. Besides, that's my poisoned apple. I'm afraid that now you're going to die at midnight."

Sandy clutched her throat and gasped for air. "Help me. I'm dying now. What's in the bag?"

"Cookies for the class party. Want one?"

"Sure," said Sandy, reaching into the bag. "I was so excited about my costume that I couldn't even eat lunch."

"I couldn't either," said Millie, taking a cookie, too.

The streets were filled with clowns, hoboes, skeletons, and ghosts, all going to Lawson School.

When they reached the playground, Millie saw another witch running toward them. Millie couldn't believe her eyes. The witch was wearing a store-bought costume—and it was exactly like Sandy's.

She spoke in Lenore's voice. "Ooh, Sandy, love your costume."

The bell rang and all the kids hurried to line up. Sandy and Lenore went up the stairs together, and Millie followed, feeling alone and invisible.

Miss Brennan was waiting in the doorway.

"Ah, the three witches from *Macbeth*," she said when they walked into the room.

"Where's *Macbeth*?" Sandy asked Millie and Lenore. Lenore shrugged. "Maybe it's somewhere near L.A."

Millie headed straight for her seat. She didn't know where Macbeth was. And she didn't care. How could they do this to her? Buy the same costume, leaving her out again?

O.C. Goodwin and Howard Hall came in right after them. O.C. was dressed as the Lone Ranger. And Howard was his faithful Indian companion, Tonto.

Marlene Kaufman came as a pilgrim. And Angela Moretti, who was dressed as a gypsy, went around reading palms. She told Millie that she was going to become rich, and told O.C. that he was going to marry Lenore Simon. O.C. made gagging sounds, and Millie couldn't help but laugh.

Not only were the kids in costume, the whole room was dressed up for Halloween. Cardboard pumpkins and jack-o'-lanterns were pinned onto the window curtains. Witches rode on brooms across the top of the blackboard, and black bats flew along with them.

Shortly after the one o'clock tardy bell rang, Miss Brennan instructed the class to line up in double file. This time Millie didn't even try to line up with Sandy. She wondered if Sandy would even want her as a partner. Probably she'd rather be with Lenore. Instead, she lined up with Rochelle Liederman, who was a box of Wheaties. Then the class passed into the hallway, down the stairs and out into the October afternoon.

All the classes in Lawson School paraded around the building. Millie knew it wasn't a real parade, with music and marching bands. It was more like a fire drill. But it was fun seeing what everyone was wearing, and it was a good way to get out of doing classwork.

As Millie walked with Rochelle, she did her best to do lots of smiling and laughing and to engage in lively conversation. She wanted to show Sandy and Lenore what a wonderful time she was having—without them.

When the parade was over, all the classes went back to their rooms. The afternoon was spent eating—popcorn balls and caramels, candy corn and Tootsie Rolls. And cookies baked by mothers and grandmothers, including the ones Mrs. Cooper had baked.

And it was spent singing—songs about witches and hats and coal-black cats. It would have been—could have been—such a perfect afternoon, Millie thought.

Millie knew she wasn't going to go to Woolworth's with Sandy and Lenore. And she told them so after school.

"You two go on without me," she said as they pushed open the heavy school doors.

"You're not coming with us?" Sandy asked.

Before Millie knew it, the words flew out of her mouth.

"I have more important things to do than go to a dime store for some stupid decorations. Besides, you can pick them out together—just like you did with the costumes." She turned and ran.

Sandy called out, "Millie, wait!"

And then Lenore: "Does that mean you're not coming to my party?"

As Millie ran, almost tripping over her long black skirt, her wart fell off and her witch's hat blew away with the wind. The hat landed across the street, right in front of the Old Colony Bottling Company.

Millie left it there and kept on running. Crazy images came to her mind: images of a witch in Millie's

black hat flying on a broomstick across the blackboard and drinking a bottle of Old Colony orange pop, of Lenore Simon wearing black-and-orange socks to match her Halloween decorations.

As Millie rounded the corner at Central Park Avenue, she glanced back to see Sandy and Lenore heading toward Roosevelt Road. Another image came to her. One of Sandy and Lenore together at tonight's party.

And not even caring that she wasn't there.

9

Still Best Friends?

The minute Millie stepped into her apartment, she changed into a fresh skirt and sweater. Off went her black blouse and the long black skirt. Gone were the hat and wart. And she even brushed her teeth to get rid of the remaining bits of licorice that were stuck to them. She didn't want to have anything to do with witches. Now she sat on the floor in front of the radio, fiddling with the knob and changing stations.

"Are you sure you don't want to go?" asked Millie's mother, sitting down next to her. "You might be missing out on a very good time."

"I don't care," said Millie. "I don't want to be at any party with Sandy and Lenore." Millie had told her mother all about their matching costumes while she was getting undressed.

"You'll have your other friends there," said Mrs. Cooper. "You'll have Angela and Marlene and Rochelle, and all the rest of the girls. Why let Sandy and Lenore spoil your good time?"

"It wouldn't be a good time with them there," said Millie, getting up from the floor. She sat down on the kitchen window seat, looking out at the leaves

69

swirling across the courtyard and thinking of Sandy and Lenore laughing together at Woolworth's. She pictured them buying balloons, and construction paper for making chains, and crepe paper for streamers, and those little paper cups for holding candies.

Every once in a while, the doorbell interrupted the picture, and Millie handed out red wax lips to the trick-or-treaters. It was almost time for supper when the bell rang again. Millie had her lips ready and opened the door.

"Trick or treat," the person on the other side called out.

"Daddy!" she cried and threw her arms around him.

"Now that's what I call a great treat."

Together they went into the kitchen for supper. But the meal was constantly being interrupted by the ringing of the doorbell. In between the visits by the trick-or-treaters, Millie's thoughts wandered to Lenore's party.

She looked at the kitchen clock. It was just after six. Probably most of the girls were at Lenore's house by now. She wondered if Sandy was there yet. Soon they would all be having fun together. And Millie . . . Millie would be answering the doorbell and handing out red wax lips.

The bell rang and Millie went to answer it. In front of her stood a witch wearing a mask and holding a pumpkin.

"Trick or treat," the witch cackled.

Millie recognized the cackle. And the costume. "Sandy?"

"Who else?" said Sandy, ripping off her mask.

"What are you doing here? Why aren't you at the party?"

"I thought we'd have our own party." She walked in and handed Millie the pumpkin. "Here. We can carve this later."

They went into the kitchen and, after Sandy said hello to Millie's mother and father, they sat down on the couch in the dining room.

"I tried to tell you something after school today," said Sandy, "but you ran away and didn't give me a chance. I tried to tell you that I didn't know Lenore was buying the same costume. Maybe it looked like we planned it, but we didn't. Her mother was supposed to make her one. It was just as much of a surprise to me."

"You didn't act like it," said Millie, looking down at her hands.

"I know," said Sandy. "And I don't blame you for being mad at me. But you know what? I've really missed you."

"I've missed you, too."

The two girls shook hands and giggled. "Still best friends, Miss Cooper?"

"Still best friends, Miss Feinman. But what about Lenore? You're good friends with her, too."

"Good friends, but not *best* friends," said Sandy.

"I mean, we do things together, but I do things with you that I would never do with her."

"Really?" said Millie. "What kinds of things?"

"Well . . . for instance, we never spill tea together. And not only that, but there are things I would tell you that I would never tell her—or anyone else."

"There are?" This was getting better all the time. "Like what?"

"Okay . . . let's see." Sandy examined the pumpkin stem. "You know how I'm always saying how much I like being home by myself while my mother and father are at work? The truth is, I hate to come home to an empty apartment. The first thing I do when I come in is turn on the radio, so there'll be voices in the house. I wish my mother stayed home like yours does."

"Oh. . . . I didn't know." Millie had always thought that Sandy liked being home alone. "I have an idea," said Millie. "You can borrow my mother whenever you want to. You can always come here."

Sandy grinned at her. "Sure. Maybe sometimes I will."

They spent the evening carving the pumpkin and listening to spooky stories on the radio. They nibbled on the pumpkin seeds after Millie's mother roasted them in the oven, and washed them down with Old Colony cream soda.

It was a wonderful party.

Just the two of them.

10

Surprise, Surprise

It was just the two of them again the next day.

"Lenore didn't come by," said Sandy, when she came to pick up Millie. They were going to see the seventh Shirley Temple movie, *The Little Princess*.

"Maybe she got tired out from the party and wants to stay home," said Millie.

"I called her house, but nobody answered."

When they reached the Lawndale Theater, they saw Lenore waiting in line with Angela and Marlene, each of them holding a purple ticket.

So it worked, thought Millie. Because of the party, Lenore got to know Angela and Marlene better, and now she doesn't have to depend on Sandy.

"I wonder why she didn't come with *us*," said Sandy. "Hey, Lenore!"

Lenore turned around. So did Angela and Marlene. Angela and Marlene waved to them. But Lenore stuck her nose up in the air and turned her back toward them.

"Uh-oh, I think she's mad at us," said Millie.

"At me, probably," said Sandy. "For not coming to her party. I had a hunch that this might happen."

The line moved into the theater, and while

Sandy went to get the Milk Duds, Millie headed down the aisle and chose two center seats in a center row.

A few rows in front of her Lenore was sitting in an aisle seat next to Angela and Marlene.

The Little Princess had some good crying scenes. Millie wiped the tears away with her hankie frequently when Shirley Temple had to say good-bye to her father, who was going off to war. Poor Shirley was forced to work as a servant in a boarding school, living in a cold, dark attic, with hardly any food.

And when Shirley finally found her father in a hospital after everyone said he had been killed, Millie and Sandy held hands and cried together. It was hard to cry and chew Milk Duds at the same time.

When the movie was over, Angela and Marlene met them in the lobby. Lenore stayed back and took long drinks at the water fountain.

"Oh, wasn't that wonderful?" Angela asked. Millie saw that she had tear streaks running down her face.

"Too bad there's just one more movie left," said Marlene.

Lenore was still drinking at the fountain.

"Is Lenore mad at us?" Millie asked.

"She's mad at both of you for not coming to her party," said Angela.

"And you missed a great party, too," Marlene told them. "Guess who showed up?"

"Who?" asked Millie, thinking of O.C. Goodwin and Howard Hall.

"You'll never believe it. Miss Brennan."

"You're kidding!" Millie and Sandy shrieked.

"She only stayed a few minutes," said Angela. "And she wasn't wearing a costume. But she looked different without an eraser in her hand, and she didn't yell at us or anything. And do you know what? I heard her telling Lenore's parents that she was never invited to a kid's party before. And she was anxious to see what one was like."

During this whole time, Lenore stayed in the background. Now she was pretending to be studying a poster of the coming attraction, *Miracle on 34th Street*.

Millie wanted her to come over and join them. She seemed so left out of everything. It wasn't fun to feel left out, Millie knew.

During the next few days, Millie and Sandy tried talking to Lenore, but she ignored them. She spent all her time with Angela and Marlene and Rochelle Liederman.

At least she's making friends with the rest of the class, thought Millie, trying to feel happy about the way things had turned out. But she wasn't happy. She missed Lenore's stories about L.A. and movie stars. And there was no one to stand up for them against O.C. Goodwin and Howard Hall.

On Friday Miss Brennan made two surprising moves. First, she stood in front of the class with her seating chart and said, "I think the time has come for

Miss Cooper and Miss Simon to change seats back. Please leave your books and just take your personal belongings."

Millie was beside herself with joy. She gathered her workbooks, her notebook with the papers spilling out, her crayon box bulging with broken crayons, and her pencil box full of pointless, eraserless pencils, and went up the aisle to her old seat. Along the way she bumped into Lenore, who didn't look any too pleased.

The second surprise came when Miss Brennan asked Millie to deliver a note to the office. Just as Sandy got up to go with her, Miss Brennan said, "Miss Simon, you may go with them."

Millie and Sandy exchanged puzzled looks as they went out the door into the hallway. Lenore followed them.

They walked in silence down the stairs to the first floor. At the foot of the stairs, Millie stopped, and in her best Miss Brennan voice asked, "How many people does it take to deliver a note? I should think that two are quite enough."

The girls stood and stared at each other, and suddenly broke out laughing. Then they sat down on the bottom steps.

"I wonder if Miss Brennan did this on purpose," Sandy said. "I wonder if she knew what was going on."

"She must have," said Millie. "Why else would she have sent the three of us? But what I don't get is how she even knew about us."

"Maybe it's because she didn't see either of you at my party," said Lenore. "I sure was mad at you guys for not being there."

"I had my reasons," said Millie.

"I know," said Lenore. "It was because of the costumes, wasn't it? I guess I should have worn a different one, like we planned. You were so mad that day I kind of figured you probably weren't coming." She turned to Sandy. "But I thought for sure *you'd* be there."

"Millie and I had some things to straighten out," Sandy answered her.

"You missed a fun party."

"We heard all about it," said Millie.

"Well, that's okay. I'm having another fun party in a few weeks. A birthday party."

"Great!" said Sandy. "We won't miss this one."

They got up from the steps and began walking down the hall. They stopped to drink at all the water fountains and peeked into the other classrooms along the way to the office. Whenever a teacher looked in their direction, they ducked out of the way and burst into fits of giggles.

"I'm glad we're friends," said Lenore. "I hope I get to stay at Lawson. I like it here."

"Why wouldn't you stay?" asked Millie.

"Well, you know about my father. He's always getting relocated. He's always being sent to different cities for his job." Lenore stopped right in the middle of the hallway.

"You can't imagine how hard it is to always be moving. To always be going to a different school. You walk into a classroom full of strangers and have all these eyes staring at you. New kids, new teachers. And just when you think you've made some friends, you have to move all over again. Sometimes it seems that it doesn't even pay to make any friends to begin with. It's just too hard."

"I hope you won't have to move again," said Sandy. "I hope you stay."

"So do I," said Millie. And she meant it, too.

When they reached the office, they stopped by the lost-and-found for a look at the latest lost items and to make sure none of the stuff was theirs.

The good old lost-and-found, thought Millie, remembering how she used to refer to it as the Lawson Found when she was in the first and second grades. A giggle rose up inside her at the memory.

"Look at this," said Lenore. "A shoe. One shoe. Some little kid in this school is walking around in just one shoe."

"Ooh," said Millie. "A Lone Ranger lunch box. Neat. I wonder if the lunch is still inside." The girls made gagging sounds.

Sandy dangled a pair of lacy underpants far in front of her. "Does this belong to anyone you know?" And the three of them burst out laughing. A dirty look from the secretary ended the laughter and the search.

Just before they went over to deliver the note, Lenore said, "You know what? All this stuff in the lost-

and-found reminded me of something. And I wondered, when you get me a present for my birthday,
could you get me something I've always wanted?
Something my mother never buys for me?"

"Sure," said Millie. "What?"

"White socks," said Lenore. "Tons and tons of
white socks."

11

New Friendships

They all sat together for *Rebecca of Sunnybrook Farm*: Millie, Sandy, Lenore, Angela, Marlene, and Rochelle. They saved the aisle seat for Lenore.

It was a happy movie, and Millie didn't have to use a hanky even once. When it was over, the girls walked out singing a song from the movie.

"Good night my friends," Millie sang, and the others joined in.

They stopped singing to wave to O.C. and Howard, who were trying to duck through the crowd again.

"You know what?" Lenore said. "For my party I'm going to invite boys."

"Boys?" the others shrieked.

"Yes. I was thinking about it and my parents say it's okay. Kids in L.A. have girl-boy parties all the time. My father says we can rent a room at the Douglas Park field house so there'll be plenty of space for everyone."

"Are you going to invite Miss Brennan again?" Marlene asked.

"Sure. And I hope she comes. Maybe she'll bring

me a present. But do you know who I really hope comes? O.C. Goodwin."

"O.C. Goodwin?" they asked in disbelief.

"He's kind of cute, don't you think?"

The other girls looked at each other. "Cute? O.C.?"

"It looks like my palm reading is coming true for you," said Angela. "I'm a pretty good fortune-teller."

Millie looked around at all her friends. My mother is a pretty good fortune-teller, too, she said to herself.

"How about coming to my house later this afternoon?" said Angela. "I've got a new Esther Williams paper doll book. She's got lots of cute bathing suits."

"I can't," said Sandy. "I have to visit my aunt."

"And I'm starting tap dancing today," said Lenore, doing a little shuffle.

"I can make it," said Millie.

"Me, too," said Marlene.

"Me, too," said Rochelle.

They walked toward Millie's apartment building because they were going to drop her off first. And as they walked, they began humming the "good night my friends" song. And then they sang the words Shirley Temple had sung to them, saying how her moment with them was now ending.

But Millie knew that the moment with her friends wasn't ending. Her friendships were just beginning.